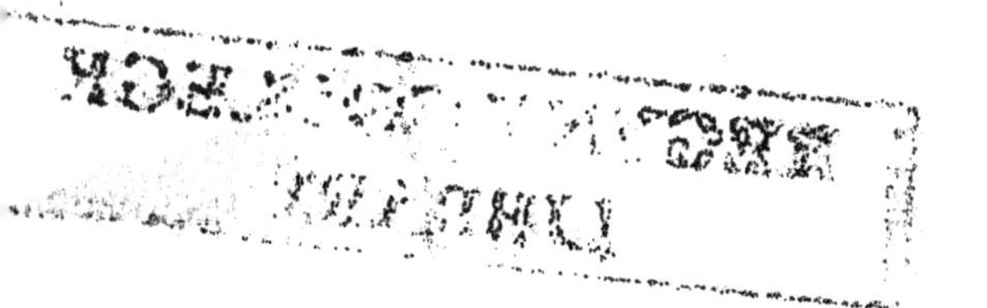

Why Science Matters

Managing Water

Richard Spilsbury

First published in Great Britain by Heinemann Library, Jordan Hill, Oxford OX2 8EJ, part of Pearson Education Limited, a company incorporated in England and Wales having its registered office at Edinburgh Gate, Harlow, Essex, CM20 2JE – Registered company number: 00872828

Editorial: Andrew Farrow, Megan Cotugno, and Harriet Milles
Design: Steven Mead and Q2A Creative Solutions
Illustrations: Gordon Hurden
Picture research: Ruth Blair
Production: Alison Parsons

Originated by Modern Age
Printed and bound in China by Leo Paper Products

ISBN 978 0 4310 4055 4
13 12 11 10 09
10 9 8 7 6 5 4 3 2 1

British Library Cataloguing-in-Publication data
Spilsbury, Richard, 1963-
Managing water. - (Why science matters)
1. Water-supply - Juvenile literature
I. Title
333.9'1
A full catalogue record for this book is available from the British Library

Acknowledgements
The publisher would like to thank the following for permission to reproduce photographs:
©Alamy **pp 32** (Alexander Johns), **35** (Frans Lemmens), **45** (Barry Lewis), **14** (Richard Osbourne/Blue Pearl Photographic); ©Corbis **pp 4, 8** (Bruno Barbier/Robert Harding World Imagery), **24** (Richard Hamilton Smith), **10** (Arne Hodalic), **33** (Andrew K/EPA), **34** (Vincent Laforet/EPA), **7** (Richard T. Nowitz), **11** (Rick Price), **20** (Tom Van Sant/Geosphere); ©Corbis/Reuters **p 31**; ©Bob Dell, www.thewaterschool.com **p 41**; ©Getty images **pp 26, 27** (Dirk Anschutz), **5** (Emmanuel Dunand/AFP), **28** (RGK Photography), **36** (Romeo Gacad/AFP); ©Jeff Godfrey **p 29**; ©NASA/JPL/The Cooperative Institute **p 19**; ©Panos Pictures/Ahikam Seri **p 42**; ©PA Photos/AP/John Flesher **p 43**; ©Photolibrary.com/Oxford Scientific/Robbie Shone **p 25**; ©Science Photo Library/Martyn F. Chillmaid **p 13**; ©Still Pictures/Ron Giling **p 39**; ©Vestergaard Frandsen/Lifestraw **p 23**. Background images supplied by ©istockphoto.

Cover photograph of the Glen Canyon dam in Arizona, USA, reproduced with permission of © Getty Images/Stone and © istockphoto.

The publishers would like to thank David Ockwell for his invaluable assistance in the preparation of this book.

Contents

Some words are printed in bold, **like this**. You can find out what they mean in the glossary.

Vital resource

There are many **resources** on Earth, such as soil and trees, but none of them is as important to human life as fresh water. Up to 66 percent of a human's body weight is made up of water. People can last for weeks without food, but they can only survive for a few days without water.

Life processes

Water is central to many life processes such as respiration and digestion. For example, the insides of an animal's lungs are wet so that oxygen from the air can dissolve in blood. Blood is mostly water. It transports oxygen from the lungs, takes nutrients to cells, and removes waste. Water also helps with temperature regulation. When humans sweat, they lose water from their bodies and cool down. Plants have roots that take in water and nutrients from the ground. Water is also a key ingredient used in **photosynthesis**, the process in which plants make their own food.

The smaller roots on a plant are covered with tiny root hairs. These increase the roots' surface area and suck up water. The water passes into larger roots, and up into the plant via a system of tubes called xylem.

THE SCIENCE YOU LEARN: WATER PRESSURE

Without water, plants wilt because they rely on water pressure to stay upright. Water pressure is the force you can feel pushing back against your hand when you hold a water-filled balloon. Unlike human cells, plant cells have a rigid cell wall composed of cellulose and vacuoles. Vacuoles are spaces that fill with watery fluid. When vacuoles are full, they push on the cell wall, and the plant cells and tissues hold their shape. Without water, the cells no longer hold their shape and the plant droops.

Managing water

Water is essential for human life, and because of this, people have always tried to manage it. Around 5,000 years ago, the first settlements were built on the banks of major rivers. People soon devised ways to lift the water out of the river, transport it to where it was needed, and store it.

Today, thanks to the work of scientists and engineers, water in most parts of the world is piped directly to homes and is safe to drink. However, for many people clean water is still unavailable, and a large number of the world's waterways are polluted. Knowledge of the science of water is essential to understanding the role water plays in the environment, our health, world industries and economies, and our way of life.

An adult loses about 1.5 litres (3 pts) of water each day through body processes such as sweating and urination. People lose even more during bouts of extreme exercise, such as running, so it is vital to replenish the body's supplies regularly.

Watering crops

To supply an ever-increasing population with food, farmers **irrigate** (water) crops to help them grow. Irrigated land provides more than 30 percent of the world's food. Irrigation is our biggest use of water. However, up to 70 percent of the world's irrigation water is wasted. Some leaks out of faulty pipes before it reaches fields, some washes away. Some rapidly **evaporates** from the soil surface as it is heated by the Sun. This means that the water turns from a liquid into a gas called water vapour and floats into the air.

Water and us

Water is pumped to our homes for washing, cooking, cleaning, and other purposes. Sanitation is the disposal of waste products, such as sewage and waste water. Using a system of pipes and clean water to carry away bodily wastes, and washing hands in clean water to rinse off germs, helps to prevent diseases spreading. Other uses of water include transport and recreation. Ships transport people and goods thousands of kilometres across the world every day. People swim in, sail on, and play in fresh water, such as lakes and rivers, all over the world.

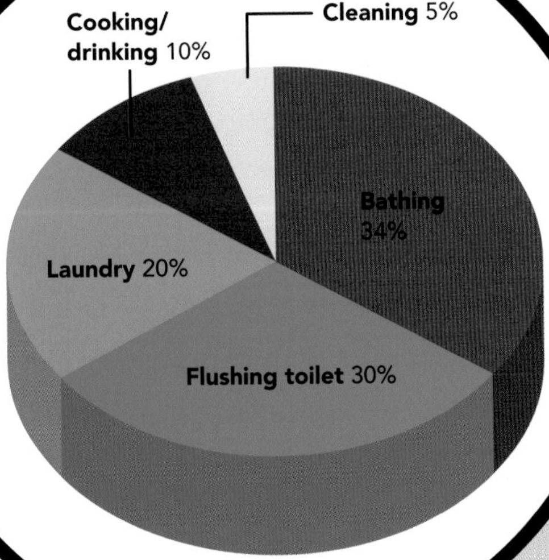

Most water is used in the bathroom. Simply turning off the tap while you are brushing your teeth can save up to 6 litres (13 pts) a minute.

Water and livestock

Farmers use water to raise farm animals such as cattle and sheep. The water is given to the animals to drink, it is used to clean the animals and their pens, it is used to grow their food, and is used to process the meat. Producing meat requires far more water than producing crops. For example, it takes almost 100 times the amount of water to produce 1 kg (2.2 lbs) of beef as it does to produce the same amount of soybeans. The high demand for meat across the world means that farmers are using more and more of our water resources for this purpose.

Drip irrigation in Zimbabwe

One way to feed water directly to plant roots is drip irrigation. A system of pipes is laid along the ground, close to the plants or just below the surface of the soil. Holes in the pipe allow small amounts of water out into the soil. The water is taken up by plant roots before it evaporates or drains away. This system reduces the quantity of water used for irrigation and reduces water wastage.

In a project in Zimbabwe, drip irrigation kits are helping poor farmers in water-scarce areas. As well as reducing water use, the kits save the countless hours spent fetching water and watering the vegetable plots. In the pilot project, 500 farmers were trained to use the technology. They increased their food production by 30 percent or more. Widow Ngwenya from Mkobokwe village says, "I am now financially independent. I can afford to feed, clothe, and pay school fees for my six children."

The water in drip irrigation pipes is filtered to stop any bits clogging up the tiny holes.

Industrial water use

Across the world, around 25 percent of all the water used by humans is taken by industry. Water is often used as a coolant in power stations and factories. Friction between a machine's moving parts creates heat that can cause damage. Water cools the machines by absorbing the heat, and carrying it away. Factories may use water as a component of their products, such as in tinned foods and fizzy drinks, or use water in the manufacturing process of their products, such as in the paper-making industry. New factories are being built all the time and by 2025 industry could be using 30 percent more water than in 2000. Much of the water used and dirtied by industries is dumped into clean water sources such as rivers.

THE SCIENCE YOU LEARN: CONSERVATION OF ENERGY

The conservation of energy is an important law of physics. It states that the total amount of energy never changes, but energy can be converted from one form into another. In a dam, potential (stored) energy is converted into kinetic (movement) energy, and then into electrical energy.

Although the total amount of energy does not change, some energy is wasted when it changes form. For example, an electric lamp may only convert 20 percent of the electricity it uses into light. The rest turns into heat energy and is lost into the air. This energy wastage drives scientists to invent more energy-efficient devices and machinery. Energy-efficient devices are less wasteful of energy.

The moving parts in giant machines that make electricity in a power station create heat. Water in pipes removes heat from the machines and the warm water cools by evaporation in cooling towers like these.

Dams trap river water in a reservoir. This raises the level of the water, which can then be released at great force through the dam.

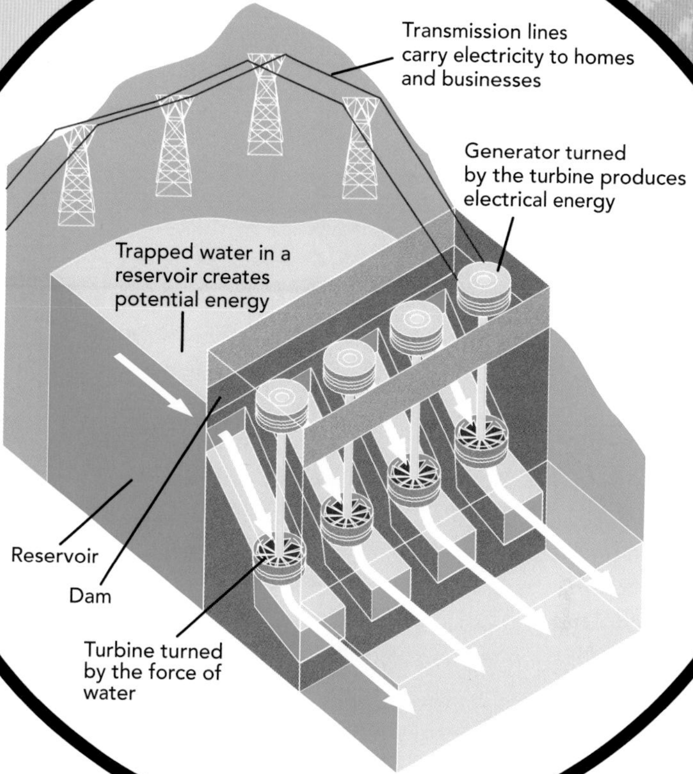

Transmission lines carry electricity to homes and businesses

Generator turned by the turbine produces electrical energy

Trapped water in a reservoir creates potential energy

Reservoir

Dam

Turbine turned by the force of water

Water power

Flowing water is used in hydroelectric schemes that provide electricity. Many hydroelectric power stations (HEPs) are built next to dams. Dams are barriers that stop the flow of river water. Behind a dam, a large reservoir of water builds up. Valves in the HEP release some of the water from the reservoir and pass it through turbines. A turbine is like a wheel that is made up of many blades. The fast-flowing water rotates the turbines at high speeds. The turbines are connected to generators. Generators convert the mechanical motion to electricity.

Hydroelectric power schemes currently generate about 20 percent of the world's electricity. The number of HEP schemes is set to increase. For example, the Three Gorges Dam on the Yangtze River in China is the biggest hydroelectric scheme in the world. When fully functional, it will provide around 11 percent of China's electricity.

Hydroelectricity is a useful **renewable** form of energy. **Fossil fuels** such as coal, oil, and gas, which are burnt to produce electricity, are non-renewable. They cannot be replaced and will run out one day. Water power is renewable because it should never run out.

Mangrove swamps provide an important wetland habitat for wildlife. They also prevent wave erosion of coastal regions.

Water habitats

Rivers, lakes, and wetlands are rich and diverse habitats. They are home to a wide range of wildlife. Some river and lake animals only spend parts of their lives in water. For example, tadpoles leave the water when they become adult frogs or toads. Other animals are adapted to permanent underwater life. A fish's adaptations include gills to take oxygen from water and fins to help it swim. There is a greater variety of animals living in the still, calm waters of lakes and ponds than there is in rivers. It is more difficult for organisms to survive in swiftly moving water currents.

Wetlands

Areas of permanently waterlogged ground are known as wetlands. In many places, wetlands have shallow waters teeming with fish. Animals that can swim but still need to breathe air, such as alligators and snakes, also inhabit wetlands. Trees and tall plants that emerge above the water are home to many insects and birds. Wetlands also provide places for migrating birds to drink, feed, and rest during their global journeys.

Ocean life

The vast oceans form the largest habitat on Earth, but wildlife here must be able to survive in salt water. Most ocean life exists in shallow waters along the coasts. The types of life found in the open ocean varies with depth. Microscopic plants called plankton float in the sunlit surface waters. These are eaten by tiny zooplankton. Together these organisms form the first link in many ocean food chains. Plants require light to make food. Therefore plants cannot survive below 200 m (655 ft) where the water is dark. Animals such as fish, squid, jellyfish, and worms can live at depths of up to 2,500 m (8,200 ft). The deepest parts of the ocean are virtually empty of life.

THE SCIENCE YOU LEARN: ADAPTATION

Physical or behavioural features that allow plants and animals to survive in a habitat are called adaptations. In a desert habitat, water is virtually absent. Cold deserts include places like the Arctic where most of the water is frozen. In hot deserts, such as the Sahara, heat rapidly evaporates the tiny quantities of rainfall. Desert organisms survive because they have water conservation adaptations. For example, some desert plants have incredibly long, shallow roots to quickly soak up any rainfall. Cacti have stems that expand to store water. Some desert animals get all the water they need from their food. Desert rodents eat seeds that look dry, but the seeds provide enough water to allow the animal to survive. Desert animals often hide in underground burrows during the day so they do not lose too much water in the heat of the Sun.

Properties of water

To understand the properties of water, we need to know some basic science. Water is made of **molecules**. Molecules are microscopically small particles. The tiniest raindrop contains billions of water molecules. Each water molecule consists of two **atoms** of hydrogen gas and one atom of oxygen. The chemical symbol for hydrogen is H and the symbol for oxygen is O. The chemical formula for water is H_2O. At room temperature, pure water is tasteless and odourless.

Water as a solvent

A **solvent** is a substance that can dissolve another substance. For example, water can dissolve salt. The water molecules separate the salt crystals into individual molecules or atoms so small that we can no longer see them. We can tell they are there because they make the water taste salty. How does this happen? Oxygen has a very small negative charge and hydrogen has a very small positive charge when they are part of a water molecule. When lots of water molecules surround a salt crystal made of sodium and chlorine, these small charges are stronger than the bonds holding the sodium and chlorine together. The charges act like a magnet, pulling the sodium and chlorine apart. This causes the salt to dissolve.

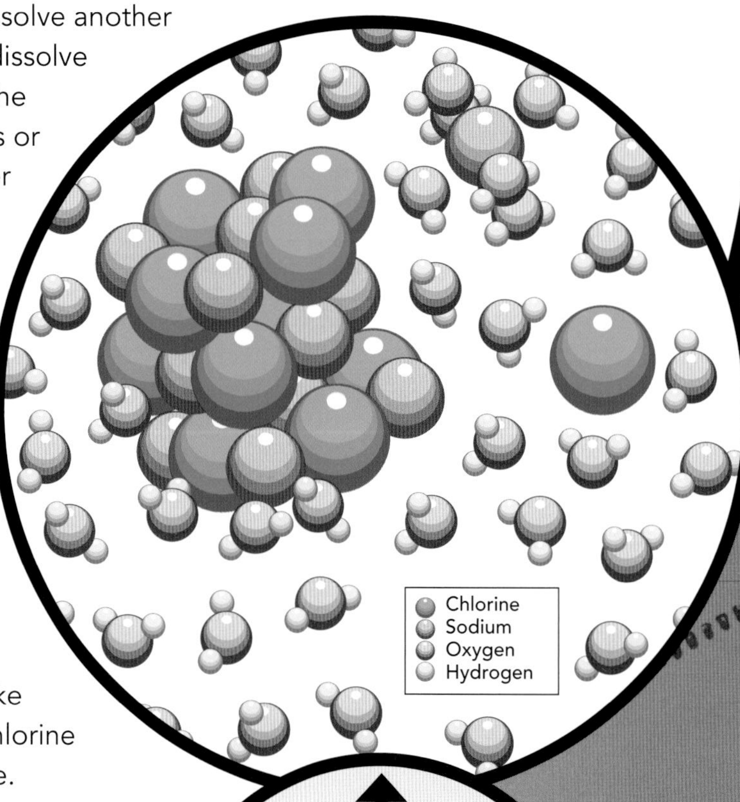

In this diagram you can see how molecules of water have broken a chlorine atom and sodium atom away from the salt crystal. The salt is starting to dissolve.

Hard or soft water?

Hard water is water that contains minerals such as calcium or magnesium. When water passes through rocks in the Earth, the minerals dissolve in the water. That is why drinking water usually has a taste. The minerals in hard water that give it flavour are also good for us. Water tastes completely different depending on where you live because the water passes through different rocks and soil.

Soft water contains few or no minerals. Rainwater is usually soft water. Soft water in water systems is treated water that contains only the mineral sodium.

THE SCIENCE YOU LEARN: SOLUTIONS

A substance that dissolves is called a **solute**. The substance it dissolves in is called a solvent. The two together make a **solution**. In a solution, the atoms and molecules of the solute are dispersed throughout the solvent. The solvent and solute in a solution can be separated. Seawater is a solution of water, salt, and other minerals. To retrieve sea salt, people create shallow pools of seawater that warm in the Sun. The heat evaporates the water, leaving the salts behind.

The photo shows a common problem caused by hard water. This section of copper water pipe is coated with a mineral deposit called limescale. Limescale can block water pipes and appliances such as kettles and irons.

Water softeners

The minerals in hard water can interfere with the cleaning ability of detergents. They react with soap or detergent to form a scum that does not dissolve in water. This effectively removes the soap or detergent and reduces its effectiveness. People living in areas of particularly hard water sometimes add water softeners such as washing soda (sodium carbonate). Water softeners react with the calcium and magnesium to form chemicals that do not react with soap. This prevents the minerals from forming a soap scum.

The three states of water

Water can exist in three states – liquid water, solid ice, and water vapour. These different forms of a single substance are called the three states of matter. The molecules in water do not change when water changes state.

Matter changes state when heat energy is added or taken away. The temperature at which a liquid starts to become a gas is its boiling point. The temperature at which a solid melts is its melting point. The Celsius scale used to measure temperature was set using water. The freezing point of water is the bottom of the Celsius scale. Therefore, water turns from liquid water into solid ice at 0°C (32°F).

Have you ever wondered how wildlife in a pond survives when the water freezes? They survive because less dense ice floats and forms a layer like a blanket over the rest of the water. This stops it from freezing.

How water changes state

In ice, the water molecules do not have very much heat energy. Heat is a form of kinetic (movement) energy. The more heat energy a substance has, the more its molecules can move about. In ice, the molecules are closely packed. They only have enough energy to vibrate back and forth. When a solid melts, the molecules are still close together but they can move about more freely. That is why liquid water flows and takes the shape of any container into which it is put. Further heating makes the molecules spread out even more. Eventually they separate from each other and become water vapour. Gases spread out to fill all available space.

Water and heat

Water has a high specific heat capacity. This means that it takes a lot of heat energy to change the temperature of water. Heat energy makes molecules vibrate and bump into each other, passing on the vibrations and heat. This is called **conduction**. In water, conduction is slow. Water takes longer to warm up or cool down than almost any other material. It also holds on to heat well. This property of water influences the climate in coastal areas. After the summer, the land loses its heat more quickly than seawater. This is why coastal regions have milder winters than places far inland. Similarly, when winter ends, seawater stays cool longer than land, so coastal places have cooler summers. This difference is known as continentality.

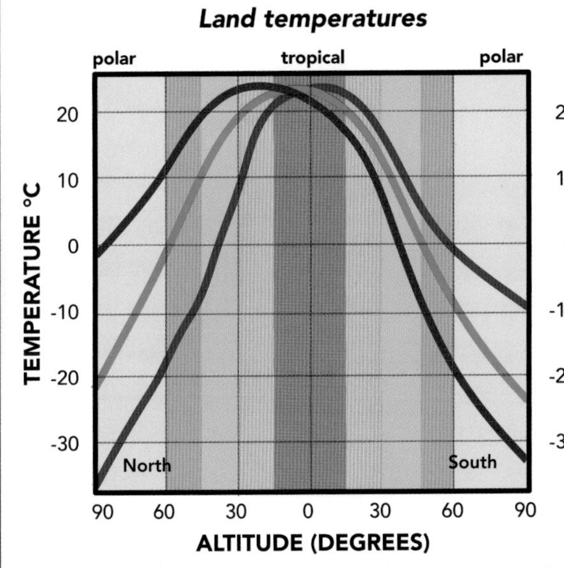

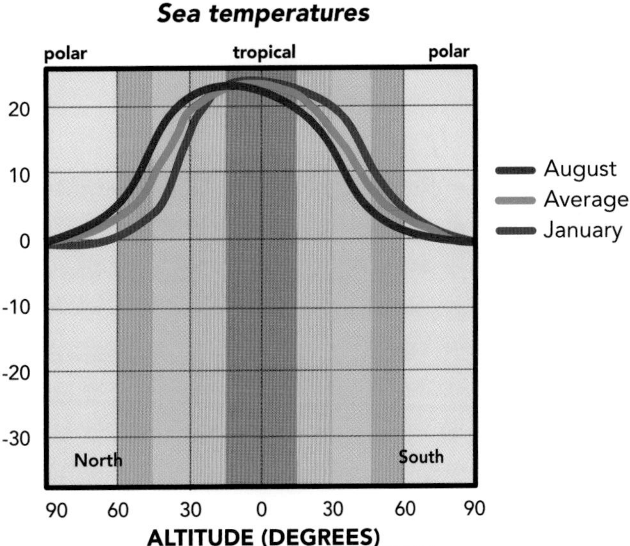

The difference between summer and winter temperatures is greater on land than at the surface of the ocean, and greater in polar regions than in tropical regions. The coldest and hottest times of year are a month later on the ocean than on land owing to water's specific heat capacity.

THE SCIENCE YOU LEARN: CONTRACTION AND EXPANSION

Most liquids contract (tighten and become smaller) as they cool and freeze. Water expands as it freezes. You can see this expansion when you freeze water in an ice cube tray. The ice rises higher than the original level of the water. Unfortunately, if water freezes in water pipes, it expands and can split the pipes and cause a leak. Water is less dense as a solid than it is as a liquid, which is why ice floats in your drink. Density is the amount of a substance in a certain volume.

The hydrological cycle

The Earth's water is endlessly recycled in the **hydrological cycle**, a process often known simply as the water cycle. A drop of water can complete one cycle in less than a day or it can take millions of years. During this time, the water changes states between liquid, gas, and solid.

Heat energy from the Sun warms water on the surface of puddles, pools, lakes, rivers, and oceans. When water is heated it evaporates and becomes water vapour. Water vapour is very light and rises up into the atmosphere.

Every 3,000 years, the hydrological cycle recycles the same volume of water as is found in all the world's oceans!

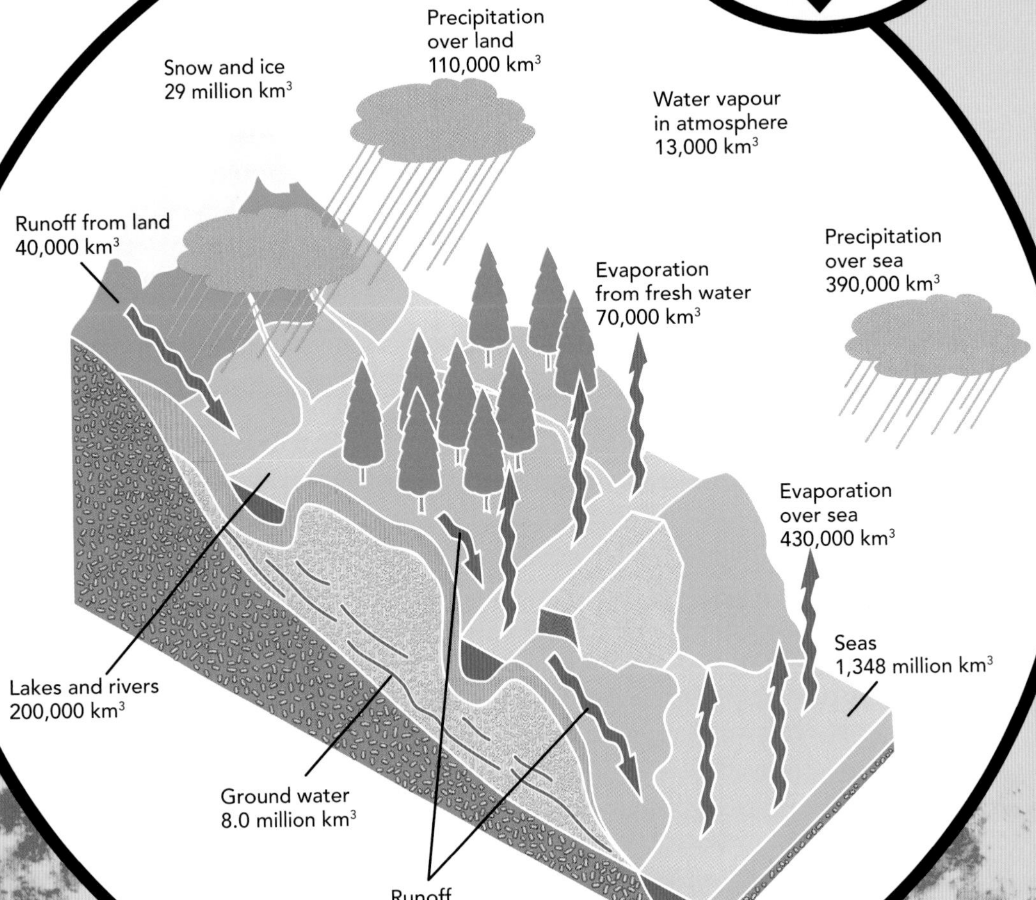

Precipitation over land 110,000 km^3

Snow and ice 29 million km^3

Water vapour in atmosphere 13,000 km^3

Runoff from land 40,000 km^3

Precipitation over sea 390,000 km^3

Evaporation from fresh water 70,000 km^3

Evaporation over sea 430,000 km^3

Seas 1,348 million km^3

Lakes and rivers 200,000 km^3

Ground water 8.0 million km^3

Runoff

The higher up you go in the Earth's atmosphere, the colder it gets. As water vapour rises, it cools. When the water vapour is cold it **condenses** (changes from gas to liquid) into tiny droplets of liquid water. Clouds form where billions of these tiny droplets of water gather together.

The tiny droplets of water in a cloud gradually join together to form larger droplets. When these droplets become heavier than air, the force of gravity makes them fall down as raindrops. If the air temperature is warmer than 0°C (32°F), clouds only contain water droplets. Droplets may freeze into ice crystals and fall as snow when the temperature cools to below 0°C (32°F). Below this temperature very cool water droplets may freeze in layers around ice crystals or dust specks in clouds. This forms hard ice balls called hail. Rain, snow, and hail are all forms of **precipitation**.

THE SCIENCE YOU LEARN: EVAPOTRANSPIRATION

Water stored in the soil eventually returns to the atmosphere via plants. Plants take in water through their roots in a similar way to sucking water through a straw. When you suck a straw, you create a vacuum and air pressure pushes water into the empty space. Water travels in much the same way through tubes from the roots to the leaves of a plant. At the leaf surface, drops of water evaporate through tiny holes called stomata, and this causes more water to be sucked in by the roots. This process is called **evapotranspiration**.

To test this you can tie a plastic bag around some of the leaves of a plant. On a plant with big leaves you may see water vapour transpired from plant leaves after about an hour. Containing it in the plastic bag makes the gas condense into drops of liquid water.

Water that falls onto mountains and hills collects in streams that run into rivers. Water that falls onto flat ground flows over **impermeable** land before soaking into **permeable** layers of soil and rock. From here it may flow into streams, rivers, and lakes. The area of land from which precipitation runs off into streams, rivers, lakes, and reservoirs is known as a **drainage basin**. Some water seeps deep underground (groundwater), where it may stay for millions of years before soaking into rivers and rejoining the cycle.

Water supply

The Sun has a huge impact on the Earth's water supply. The position of the Sun is also the reason water exists in different states on Earth.

Light from the Sun travels to Earth in straight lines, called rays. The Earth's surface is curved, which means the rays hit the ground at different angles. The area around the Equator (the imaginary line around the centre of the Earth) is the closest part of Earth to the Sun. Here, the rays hit the Earth straight-on, which makes it the hottest region of the Earth. To reach the North and South Poles, the rays travel further and lose some of their warmth on the way. Also, they hit the ground at an angle so they cover a larger area of the Earth's surface. The Poles are the coldest places on Earth and most water exists here as solid ice.

Clouds and the weather

Water vapour in the atmosphere is responsible for most of the weather we experience. The moisture in the air creates clouds, which bring precipitation, thunder and lightning, fog, sea mist, and severe events such as hurricanes. Cloud cover also affects temperature. When the Sun's rays hit the Earth's surface, they bounce back towards space. The water vapour in clouds acts as a shield and traps heat in the atmosphere.

It rains regularly close to the Equator because air warmed by the Sun holds a lot of water. At the Poles, the air is so cold that water only falls as snow.

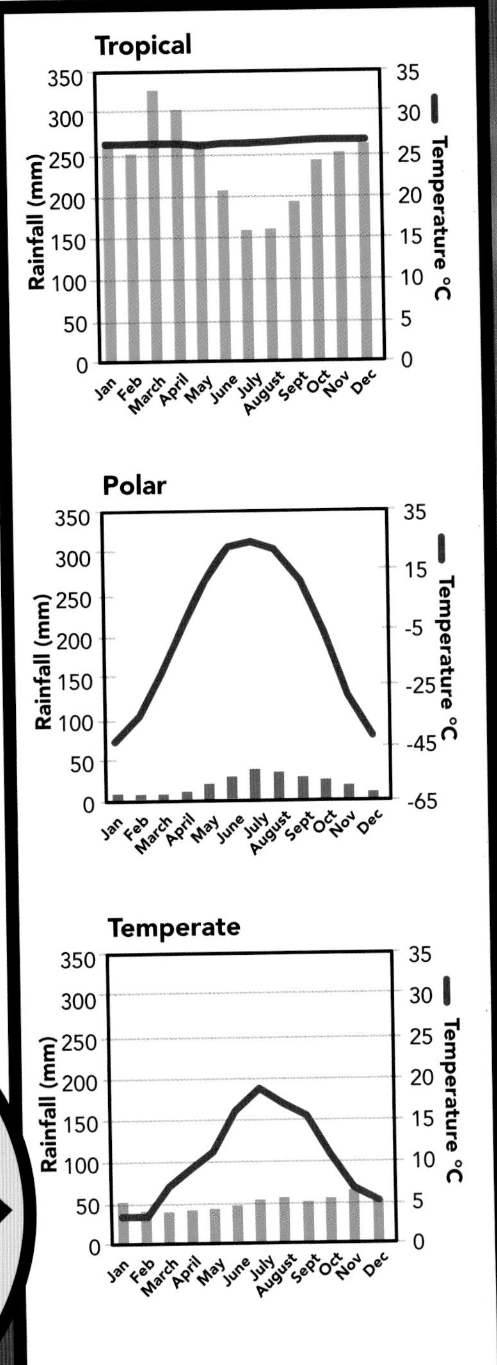

The shape and size of a cloud depends on factors such as **humidity**, temperature, air motion, and **aerosols**. Aerosols are tiny airborne particles such as dust and pollen. Water vapour tends to condense on to aerosols, so they are known as condensation nuclei. Areas of atmosphere with lots of aerosols form many clouds.

CUTTING EDGE: SEEING THROUGH CLOUDS

In 2006, a team led by NASA scientists launched two new satellites to obtain three-dimensional images of clouds. The CloudSat satellites collect data about the vertical structure of clouds, the amount of liquid water and ice in them, and how clouds affect the amount of sunlight that passes through.

Each satellite transmits pulses of energy towards the clouds and measures the number of pulses scattered back. This allows CloudSat to build up images of the clouds. The satellites can also distinguish between cloud particles and precipitation. The measurements will help scientists study and predict weather patterns. "CloudSat's radar provided breathtaking new views of the weather on our planet," said Graeme Stephens, CloudSat principal investigator. "All major cloud system types were observed, and the radar demonstrated its ability to penetrate through almost all but the heaviest rainfall."

This is a CloudSat image of a cross-section through a vast storm cloud in the Caribbean in August 2007. The mauve and red parts of the cloud are those that reflect the radar beam the most. This means they are full of water droplets and ice crystals. The blue parts reflect the least and have less moisture inside.

This satellite picture shows Antarctica and its surrounding waters. Vast areas of water around this continent are unavailable to us, either because they are salty or locked up in ice.

Freshwater supplies

Earth is commonly referred to as the blue planet because water covers almost 75 percent of its surface. About 97.5 percent of this water is salty seawater. For millions of years, rainwater has passed through or over soil and rocks before flowing into the seas and oceans. As it does so it dissolves small amounts of mineral salts from rocks. This salt has built up in the oceans, making them salty. Less than 2.5 percent of the world's water is fresh water, and most of this is either underground or frozen into ice in glaciers and ice sheets. Less than one percent of the blue planet's water is both fresh and available for us to use.

Groundwater stores

Water seeps into the ground through rocks and soil in a process known as **infiltration**. The more cracks, holes, or openings in the Earth's surface, the more infiltration occurs. Some of this groundwater collects in underground pools and streams where it may stay for a long time. Some groundwater soaks into layers of porous rock (rock that contains many tiny holes). Rock that is **saturated** with water in this way is known as an **aquifer**. Wells that reach below the **water table** are drilled to access groundwater. The water table is the top layer of porous rock that is saturated with groundwater.

Surface water stores

Surface water sources include rivers, lakes, ponds, and reservoirs. Rivers form naturally as water from rain or melted snow flows down over sloping land until it joins the sea or flows into lakes. Lakes and ponds form when rivers and streams flow into and enlarge natural dips or hollows in the Earth's surface. Reservoirs are artificial lakes. They are either formed by dams, or purposely dug out and lined with concrete to become watertight stores for fresh water from rivers and rainfall. Surface freshwater sources like these provide a large percentage of the world's freshwater needs.

Water distribution and treatment

Water managers plan the construction of dams, reservoirs, and pipelines to ensure there are sufficient supplies for ever-growing communities. They make sure that sewage, water treatment, and distribution systems function properly. Waste water and sewage is collected and treated before being released into rivers and seas. Water collected for distribution to homes or factories is sterilized for drinking. At the beginning of the 20th century, scientists discovered that killer diseases, such as cholera and typhoid, were caused by microorganisms in water. They started to add a chemical called chlorine to the water to kill germs and make the water sterile. Today, ultraviolet (UV) light and other treatments are also commonly used.

INVESTIGATION: BUILDING A WATER FILTER

You can create a filter to clean impurities from dirty water using simple materials. [**WARNING**: After this experiment, the water will have been cleaned, but not sterilized. You should not drink the water.]

Equipment

- scissors
- two clear plastic cups that can stack on top of each other
- 10 cotton wool balls
- a square of plastic screen mesh
- sand
- gravel
- two beakers
- dirty water from a pond, puddle, or stream
- rubber gloves

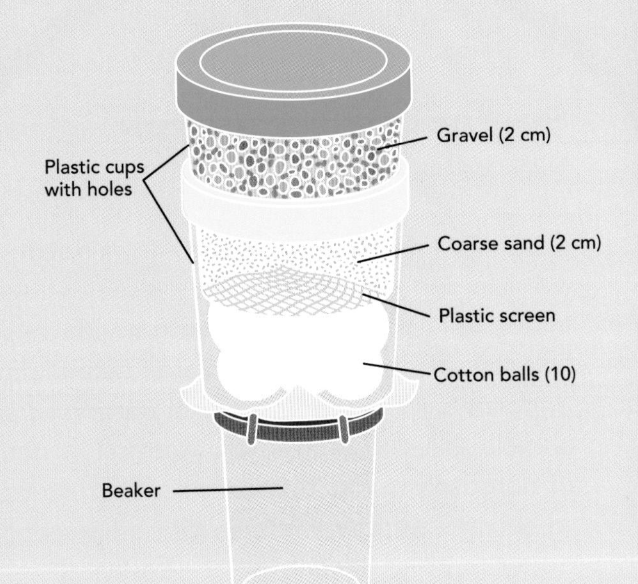

Procedure

1. Use the scissors to carefully pierce several holes in the bottom of each plastic cup.
2. Place the cotton wool balls in one cup and the mesh in the other.
3. Stack the cup with the mesh on top of the cup with the cotton wool balls.
4. Pour sand to a depth of 2 cm (0.8 in) on the mesh and then add the same depth again of gravel.
5. Half fill one of the beakers with dirty water. You could collect this from a pond, puddle, or stream, but make sure you wear rubber gloves when you do so.
6. Put the filter unit you made using the two plastic cups on top of the second beaker.
7. Tip the dirty water onto the gravel. The water that comes through into the second beaker should look clearer.
8. Use the filter many times. The cotton wool will start to look darker from the dirt it removes from water passing through it.

The LifeStraw® filter

Half of the world's poor suffer from waterborne disease as a result of drinking unsafe water. These diseases kill around two million people each year. Filtration alone does not remove all the microorganisms causing waterborne diseases from dirty water. However, sucking dirty water through a LifeStraw® (see photo) takes it through filters and also chemicals that kill dangerous microorganisms. A person can drink a litre through the straw in eight minutes. Each LifeStraw® can purify 700 litres (185 gallons) of water – a year's worth of drinking water for one person. LifeStraw® inventor, Danish scientist, Torben Vestergaard Frandsen, came up with the idea while developing a filter to remove guinea worm larvae from water. "The question then came up that if we can filter out one parasite, why not also filter out all bacteria and viruses at the same time? The real spur for this was the realization that you can create the necessary force for the filtration with your personal sucking power." Although the LifeStraw® only costs around £2 ($U.S.3), some charities say this is still too much for most people in less economically developed countries (**LEDC**s). They also say that the real issue is that water supplies are too far away from people's homes.

Shaping the planet

Our landscape is shaped by **weathering** and **erosion**. Weathering happens when natural forces, such as water, temperature, and wind, break down rocks on the Earth's surface. Erosion is when the rock pieces are washed or blown away. Soft rocks such as clay and sandstone are more easily weathered and eroded than harder rocks such as granite.

Physical weathering

Rock surfaces are broken down by physical weathering. Freeze-thaw is a type of physical weathering that can occur when water seeps into cracks in rock. When the water freezes, it expands and enlarges the cracks. When the ice thaws, more water fills the larger cracks. When this process happens repeatedly, the surface layer of the rock eventually loosens and breaks off.

Over time, the freeze-thaw cycle forms cracks in solid rock and even splits it into smaller pieces.

Chemical weathering

Rock can also undergo chemical weathering. Carbon dioxide in the atmosphere dissolves in rainwater. This creates very weak carbonic acid. When the acidic rainwater trickles through the ground, it can dissolve calcium carbonate in chalky rocks such as limestone.

Acid rain

Human activity can increase chemical weathering by creating **acid rain**. Burning fossil fuels in power stations and in vehicles releases acidic gases such as sulphur dioxide. In the atmosphere, water vapour condenses onto these gases and dissolves them, forming an acid. When these water droplets fall to Earth they are known as acid rain.

Acid rain can contaminate drinking water and lakes, which kills fish and other aquatic life. Acid rain ruins forests. It damages trees and other plants by affecting their ability to take in nutrients through their roots. It can also dissolve rocks and stonework on some buildings.

THE SCIENCE YOU LEARN: STALACTITES AND STALAGMITES

The mineral solution created by acidic rainwater dissolving calcium carbonate will sometimes seep through cracks in the ceilings of limestone caves. As the solution drips onto the floor, the water very slowly evaporates, and the calcium carbonate is left behind. This accumulates over thousands of years to form stalactites and stalagmites. Stalactites hang down from cave ceilings and stalagmites stick up from cave floors. These structures develop very slowly – some stalactites can grow as slowly as 1 cm (0.4 in) every 1,000 years!

River erosion

Rivers shape the land by wearing away rocks, then transporting and depositing (dropping) rock debris and **silt**. In the upper course (stage) of a river, the gradient is steep. Water running down hills or mountains has lots of energy. The river erodes vertically. It carves deep but narrow channels, such as the mighty Grand Canyon in the United States, which was shaped by the Colorado River.

In the middle course of a river, the gradient is less steep. The river weathers and erodes laterally (to the sides), so river channels and valleys tend to be wider and follow a more winding path. In the lower course of a river, there is little or no gradient. The slow-moving water no longer has the energy to carry along **sediment**. When rivers deposit sediment in a region where there are no tides, a delta forms. A delta is an area of land formed from sediment. Deltas are often fan-shaped, such as you find at the mouth of the River Nile.

Flood plains

A flood plain is the land along a river that is covered by water during a flood. The deposit of sediments on a flood plain makes the soil fertile. This is one reason why people often build farms on flood plains. However, floods can wash away homes and kill people and livestock. In many places people are discouraged from living on flood plains, but this is not always possible. Some people are too poor to buy land elsewhere. In 2007, floods in Bangladesh and northern India made more than 1.5 million people homeless. (Find out how scientists are helping to save people from floods on pages 34–35.)

The powerful force of river erosion has removed rock from a wide area of mountainside, leaving a grooved valley. The land here was as high as the tall mountains in the background before erosion began in the past.

A river winds through flat plains. The grey silt by the water was once rock from higher ground. It was eroded and deposited by the river.

CASE STUDY

Amazon dynamics

The River Amazon is the world's widest and deepest river. It contains 60 times more water than the River Nile. It runs for 7,100 km (4,410 miles) from its source in Peru 6,500 m (21,330 ft) up in the Andes Mountains down to sea level at the Atlantic coast. The Amazon drainage basin covers the entire central and eastern region of South America.

In the upper stages of the River Amazon, the raging waters cause extreme erosion, and transport millions of tonnes of sediment every year. Erosion in the upper reaches of one of its tributaries is more than 3,000 tonnes per sq km (8,700 tons per sq mile) of land per year. Most of this never reaches the sea. Instead, around 60 percent is deposited in the foothills of the Andes mountains.

Ocean erosion and deposition

Wave power erodes shorelines. Erosion at the base of a cliff can cause the top of that cliff to collapse. Where there are areas of hard and soft rock, the harder rock may be left jutting out as headlands. When a crack forms along a headland, the seawater may eventually open this crack and form a cave. After further erosion, the cave can open and collapse, leaving behind an arch or stack (a column of rock).

Pieces of rock eroded from the coast smash into smaller pieces, forming pebbles and sand. Ocean waves deposit the pieces elsewhere, creating beaches between headlands and on flat regions of the shoreline. Waves also move material along the coast. **Longshore drift** can drag sand and pebbles sideways along a coast, creating coastal features such as **spits**.

Managing coastal erosion

Coastal erosion and instability can put coastal towns and villages at risk. **Geoscientists** monitor how coasts are changing and advise coastal managers which parts of the coast should be defended and which kind of defence to use. Traditional defence methods include sea walls and groynes. Sea walls deflect waves away from buildings. Groynes are like wooden fences placed at right angles to the coast to reduce longshore drift. Such sea defences can be expensive and may cause coastal erosion on stretches of nearby coast.

Natural solutions to coastal erosion include planting more vegetation or establishing sand dunes. Planting marsh or reed beds in an estuary can bind sediment together and slow erosion. These measures protect coasts from erosion, storm surges, and coastal flooding.

Marram grass is often planted to create sand dunes. This plant grows happily in sand and its root system binds the sand together.

Scientist at sea!

With funding from NASA, scientist Heidi Dierssen (see below) is studying how space satellites can be used to monitor seagrass in shallow coastal waters. Seagrass grows on the ocean floor and as well as being a wildlife habitat, it slows down waves and so protects shorelines from erosion. Unfortunately, seagrass has been decreasing due to coastal development, water recreation, and invasive species (non-native organisms).

Satellite observations of ocean colour can be used to work out concentrations of underwater plants, but satellites have difficulty seeing through to the ocean floor. Dierssen dives under the sea and uses sensors to measure the amount of light passing through the water and reflecting off the ocean floor. She uses this information to develop mathematic formulas that help scientists work out what is on the ocean floor using satellite images of ocean colour. This allows them to track the progress of seagrass restoration projects.

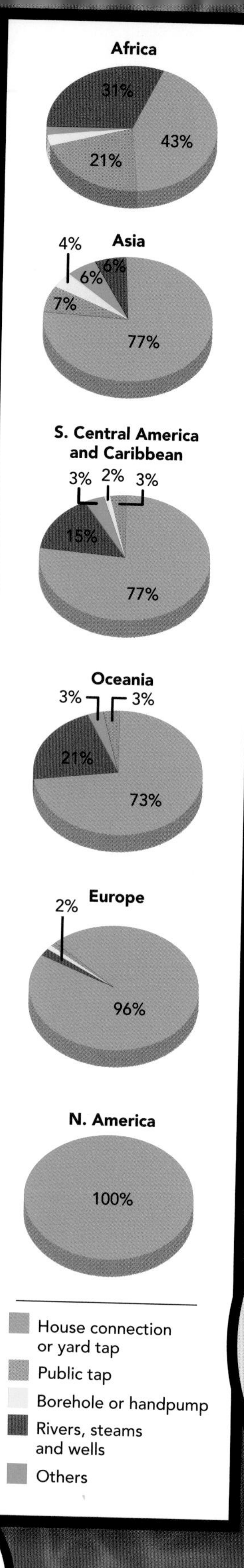

Changing water availability

Water is a vital resource and yet around 30 percent of the world's population do not have an adequate supply of clean, fresh water. In some hot, dry countries there is simply very little rain. Hot, dry countries may also suffer periods of drought, when there is so little rainfall for so long that the soil dries out and plants die. In parts of Somalia, Africa, droughts have lasted for years on end. In parts of Asia, although the annual rainfall is adequate, most of the rain falls in short, heavy bursts during the monsoon season. On average, India receives 80 percent of its annual rainfall in just 200 hours and 80 percent of this rain simply washes out to sea.

Population factors

The rapidly expanding world population is a major pressure on water availability. In countries such as Pakistan and Nigeria, populations are expected to triple by 2050. Not only do more people need more water, but the amount of water people use has doubled in the last 30 years. Rates of useage are still increasing. When standards of living increase, more people across the world benefit from improved sanitation. They start to rely on washing machines, showers, baths, and flushing toilets. The amount of water used per person increases dramatically.

These charts show the water availability in the largest cities on each of the world's continents. They also show the percentage of the population with each type of service.

People gather to draw water from a huge well in a village in the state of Gujarat in western India. At the time this photo was taken, the region was experiencing the worst drought for over a decade.

Access to managed water supplies

In some places there is sufficient water but it is not easily accessible. In more economically developed countries (**MEDCs**) and in most cities across the world, water is piped straight to people's homes via a water distribution network. In LEDCs, people in rural areas may have to walk long distances to collect water from a river or a communal well, and this limits the amount they use. For example, each day someone with water piped to their home may use up to 1,000 litres (2,110 pts), while someone who collects water from a river 2.5 km (1.5 miles) away uses less than 5 litres (10 pts).

THE SCIENCE YOU LEARN: LOW-FLUSH TOILETS

Low-flush toilets help save water. A conventional toilet uses an average of 13 litres (27 pts) per flush, while a low-flush toilet uses only 6 litres (12.5 pts) per flush. You can also save water by using a water displacement device. Put a brick or a water-filled one-litre (33-oz) plastic bottle inside the toilet cistern. This reduces the amount of water the cistern holds and so reduces the amount of water that is flushed away.

Draining and drying wetlands

People control or manage most of the world's rivers and wetlands in some way. Many wetlands and delta regions have been and continue to be drained so the dry land can be used for farms or for building homes and factories. Around half of the world's original wetlands were altered in this way during the 20th century. For example, although the huge Coto Donana wetland in Spain is protected, it is still under threat. Excessive amounts of its groundwater are taken for agriculture and to provide water for tourist developments.

As well as providing water and homes to many of the world's animals, wetlands act like sponges, soaking up water and then releasing it slowly. This means they are good water storage facilities, and they help to prevent flooding and erosion, especially in coastal regions. Wetlands also help to keep other freshwater resources healthy. When water passes through the plants, the roots filter silt and sediment from the water and make it clean. This is so effective that in some places people have constructed artificial wetlands to treat dirty water.

River diversion

Many rivers are diverted away from their natural course to provide irrigation. River flow is also affected by dams built to create reservoirs for hydroelectric power plants. Dams are sometimes built to reduce or control water flow in order to prevent flooding downstream.

Large parts of the Everglades wetlands in Florida have been drained, forcing wild animals into closer contact with humans.

This type of water management can also cause problems. Reduced water flow means there is less water for people downstream to use, and fewer fish for them to catch. Rivers slow down and deposit their sediment in the reservoir instead of further downstream on the flood plain. The reservoir can silt up. The flood plain becomes less fertile and less useful for growing crops such as rice. When reservoirs are built, large areas of land are flooded and people are forced to move from the area.

CASE STUDY

The Three Gorges Dam

The Three Gorges Dam on the Yangtze River in China (see below) is the largest in the world. It was built to make hydroelectric power for millions of people, and prevent flooding downstream. Big dams like this are a controversial solution to power and water management and the Chinese government has admitted there are problems. The Three Gorges reservoir has submerged ancient temples, towns, and villages, and has displaced more than one million people. Landslides caused by erosion on the steep hills around the dam have deposited thousands of tonnes of sediment, affecting the HEP turbines. There have also been disputes over land shortages, because farmers forced to move to higher ground share small plots. The quality of some of the region's drinking water has deteriorated and pollution is seeping into the reservoir from industrial sites submerged when the dam raised the water level.

Flood disasters

Floods occur when rivers burst their banks and water flows onto a surrounding flood plain. Floods are usually triggered by heavy rainfall. Floodwater can move with great force and speed, and can be extremely destructive. Settlements are destroyed, and people and animals drowned. Flood water becomes polluted with mud, sewage, and other waste so that there are no supplies of fresh water.

Flood management techniques include restricting the amount of building allowed on flood plains, building dams to control water flow, and changing a river's shape. New channels that branch off the main river can remove excess water, and levees can be created. Levees are high banks made of earth or concrete, built to prevent rivers or lakes from flooding. However, none of these methods is guaranteed. In 2005, the force of Hurricane Katrina caused two giant levees to burst in New Orleans in the United States. Water engulfed the city.

In 2005, Hurricane Katrina caused terrible flooding in New Orleans, USA. Heavy rain and surges of water broke through levees built to prevent the Mississippi River and a nearby lake from flooding. More than 700 people died as a result of this disaster.

Landslides

Landslides may happen where mountain forests have been cut down. Trees absorb some of the force of heavy rain, and tree roots hold topsoil together. Without trees, water rushes straight down the slopes instead of seeping into soil and entering streams and rivers. The soil turns to mud that is carried down the slopes, dislodging rocks. Landslides can bury entire villages. Landslide prevention techniques include reforesting hillsides and building defences, such as tough metal barriers, to hold back mud and rocks.

Dutch flood expertise

The Netherlands is a country partially built on wetlands and criss-crossed with waterways. Forty percent of its land area has been reclaimed from the water, and Amsterdam's Schiphol Airport sits on what used to be a lake. Without careful management, more than 60 percent of the country would be regularly flooded!

The Netherlands prevents floods in several ways. Just off the port of Rotterdam there are two massive, computer-controlled storm surge barriers (see the photo above). These curved gates swing shut in the event of extremely high water, as happened in autumn 2007. Barriers have been built along the coast, sea inlets have been closed off, and vast areas of sand dunes developed. There are dikes (banks of earth) to hold in water, and land designated for rivers to flood when the water level rises. In this way, the soil and vegetation can absorb rainwater and slowly discharge it back into the river network.

Water pollution

Poisons, chemicals, and other substances contaminate water sources and cause pollution. This reduces the amount of clean water available for people to use. Polluted water is expensive to purify and some types of pollution are impossible to remove. Some water pollution results from natural causes, for example when volcanic gases dissolve in lakes, but most is a result of human activities.

In 1996, a drainage pipe from the Marcopper Mine in the Philippines (producing copper, silver, and gold) burst. Between 3 and 4 million tonnes (3.3–4.4 million tons) of mining waste polluted the Boac River.

THE SCIENCE YOU LEARN: FOOD WEBS

Pollution in a habitat can be passed on through **food webs**. For instance, ocean currents wash pollutants from rivers in Europe to the Arctic. The pollutants linger in the cold Arctic waters and contaminate plankton (the microscopic plants and animals that float near the surface). Plankton are eaten by fish, which are themselves eaten by larger fish, birds, seals, and then polar bears. The concentration of pollutants multiplies five to ten times with every step up the food web. This **bioaccumulation** puts animals at the top of the Arctic food web at the highest risk of health problems, such as stunted growth and damage to their reproductive systems.

Industrial pollution

Despite laws banning the dumping of waste into water sources, millions of tonnes of industrial waste get into surface and groundwater supplies every year. Factories sometimes dispose of waste metals, solvents, or other chemical, on land or in water. Some mines regularly dispose of toxic (poisonous) waste in rivers, lakes, and even oceans.

Agricultural impacts

Chemical fertilizers and pesticides are used to increase crop yields, but they can pollute water when they run off fields into streams and groundwater sources. They can also cause **eutrophication**. Nitrogen-based fertilizers work by adding nutrients to the soil. When these nutrients flow into water sources, plant-like algae feed on them, and grow and multiply quickly until they cover areas of water in green slimy blankets. These algal blooms effectively suffocate the habitat and can even kill wildlife.

In oceans, eutrophication can create dead zones – areas of water so low in oxygen they cannot support marine life. In the Gulf of Mexico there have been dead zones like this for 30 years, some covering areas of up to 17,000 sq km (6,500 sq miles). They have been caused by fertilizers flowing down rivers such as the Mississippi and into the Gulf.

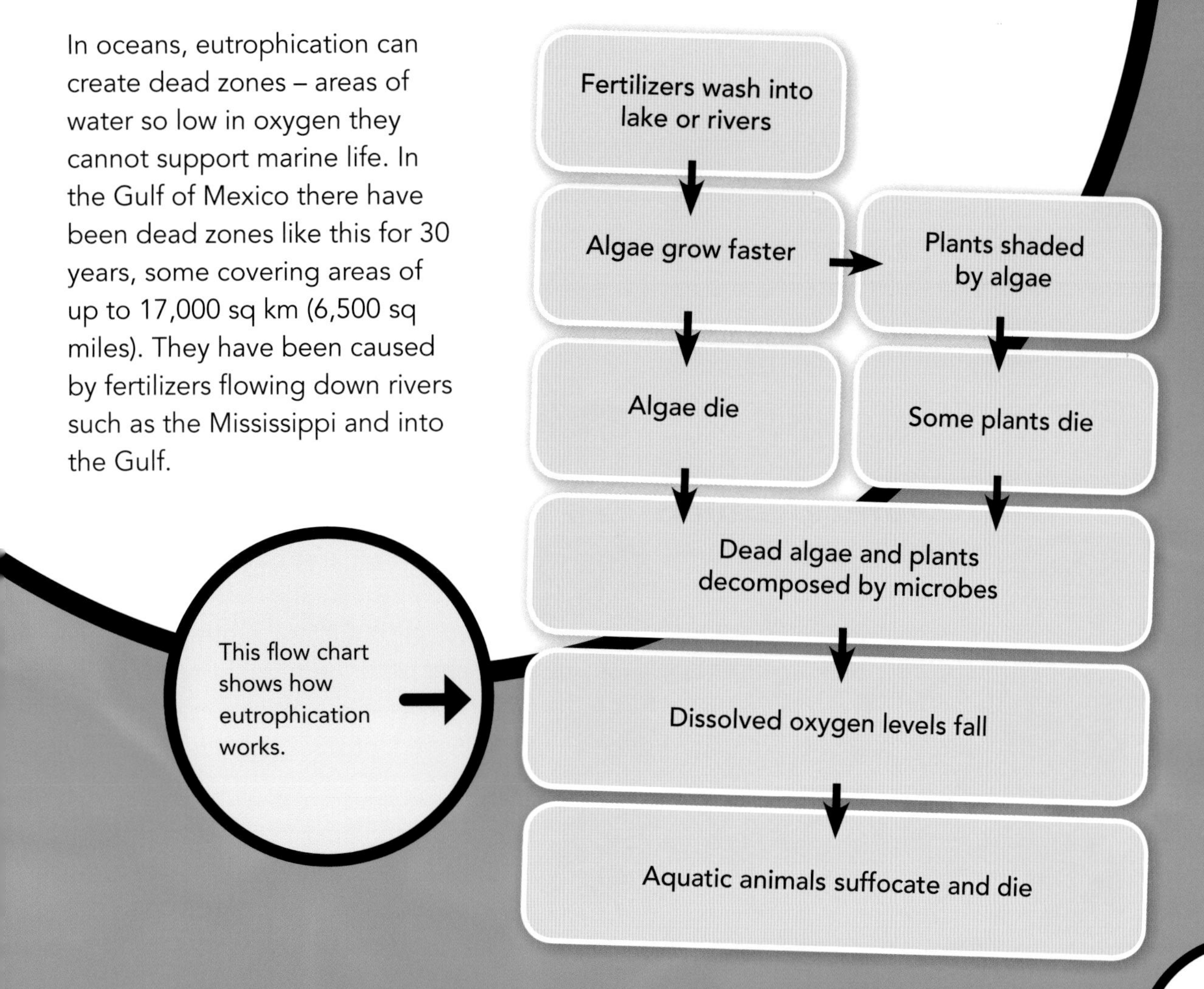

This flow chart shows how eutrophication works.

Sewage and sanitation

Sewage is the major cause of pollution in many countries of the world. An estimated 90 percent of wastewater in LEDCs is still released directly into rivers and streams without any treatment. More than 30 percent of the world's population live without improved sanitation. This means that nothing is done to treat and purify their sewage before it is discharged into rivers, lakes, or the ocean. People do not have flush toilets or even private pour-flush latrines (toilets that wash waste into contained pits using buckets of water). This is not ideal, but sometimes health officials have to make difficult choices. Piping waste into a river may be the only alternative to leaving raw sewage near people's homes, where it would be an even greater health hazard.

Managing sewage systems

Even in MEDCs with improved sanitation, wastewater can get into water sources. During storms, rainwater may enter the sewage system through faults in pipes, causing overflows. Sewage can also overflow into water supplies when pipes are cracked or blocked by tree roots or other materials. This happens because sewage systems have overflow points. These are valves that open if a problem occurs, and release sewage to prevent it backing up into people's homes. During unusually wet weather in 2004, more than 12 million cubic metres (3.1 billion gallons) of raw sewage was discharged into the River Thames from five overloaded pumping stations. That is the same volume as almost 5,000 Olympic-sized swimming pools. Thousands of fish were killed.

Wastewater problems

Sewage contains bacteria that pose risks to human health. It also contains nutrients from food that people have eaten, which can lead to algal blooms and eutrophication. Other household wastewater contains dirt and synthetic (artificial) chemicals from shampoos, shower gels, and other detergents. Many of the chemicals used in household cleaners and cosmetics are strong enough to survive water purification treatments and are released into the sea. They are believed to be harmful to living things.

Sanitation in Mumbai's slums

Today, 75 percent of the population of LEDCs live in slums. Slums are areas of houses built from materials such as wood pieces, card, corrugated iron, and broken bricks.

In Mumbai, India, almost 7 million of the 12 million city residents live in slums. In the Ganesh Murthy Nagar slum in the Colaba district of Mumbai, one resident said, "We had one small, smelly toilet for a population of 10,000. Women suffered the most because they had to relieve themselves in the open, and could do so only in the early mornings or after dark." Five million Mumbai slum dwellers live without toilets and around 2.5 million kg (5.5 million lbs) of human waste contaminates their environment each day. This leaves the local population at serious risk from infection and disease.

The reality of daily life for many poor people around the world is an overpopulated slum with open sewers choked with waste.

Waterborne diseases

More than two million people die every year from diseases linked to polluted water. Around 90 percent of those who die are children. Most of these deaths could be prevented if people had access to clean water and sewage systems. Cholera and typhoid are waterborne diseases caught by drinking water contaminated with human or animal faeces (excrement). People are also killed by other water pollutants. For example, arsenic is a naturally occurring poison that contaminates the drinking water of millions of people worldwide. It causes a slow illness that often ends in death.

Parasites and disease

Some waterborne diseases are caused by insects and other animals that develop or breed in water. Mosquitoes are water-breeders that carry disease. Covering standing water, emptying out barrels regularly, and treating water with insecticide can prevent mosquitoes from breeding.

Sometimes, water management can cause problems. Schistosomiasis (or bilharzia) is a disease caused by a **parasite** carried by freshwater snails. It makes people very sick, and can cause bladder cancer and kidney failure. In Senegal, Africa, a dam built along the Senegal River created the perfect habitat for the snails and introduced schistosomiasis into both Mauritania and Senegal.

This chart shows the world distribution of deaths due to dirty water for the year 2000. In 2007, the World Health Organization estimated that 1.5 million children under five years old were killed by waterborne diseases.

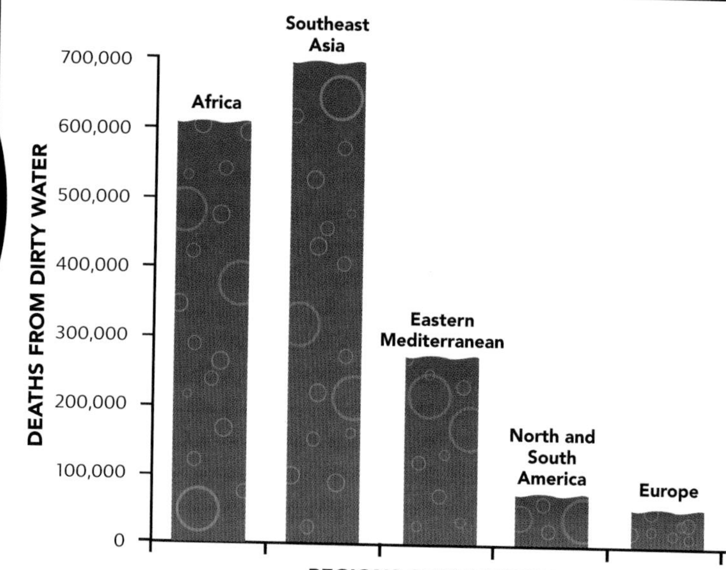

CASE STUDY

SONO filters

In 2007, Bangladeshi chemistry professor Abul Hussam won an important prize for inventing a cheap, simple system for filtering arsenic from drinking water. Hussam's filter is designed to be used at community or residential wells, and uses a two-stage system. Water first passes through a plastic bucket filled with one layer of coarse river sand and one layer of cast iron material. The sand removes sediment, and the iron traps soluble arsenic compounds. The second stage is a bucket containing a layer of coarse sand, a layer of wood charcoal, and a layer of fine sand and brick chips. The charcoal absorbs more arsenic, while the sand and brick chips remove finer particles. These SONO filters have already saved thousands of people from health problems.

Solar water disinfection uses the Sun's heat to kill bacteria that cause diseases. Clean, transparent plastic bottles are filled with water from wells, the lids are screwed on, and they are left in the Sun for six hours. This is cheaper than boiling water on a stove and more environmentally friendly.

Water solutions

By 2050 it is estimated that nearly half the world's population will be severely short of clean, safe water. Many people consider world water management to be the most important challenge that faces the globe today.

Solutions with problems

Solutions to water shortages can create new problems. For example, in some regions people have dug deeper underground to reach new groundwater sources. Many aquifers formed thousands of years ago when the climate was wetter than it is today. Aquifers take so long to replenish that once used up they are effectively no longer a freshwater resource.

Over-use of groundwater can also cause salination. This is when dissolved salts contaminate water. In particular, salination can happen near coasts. When underground aquifers are pumped dry, seawater soaks in to the area to replace it. Sometimes when aquifers are drained, salts dissolved in deep soils are brought to the surface. When the water evaporates, the salts remain. They dry out or kill plants and leave soils infertile, which means the land can no longer be used to grow crops.

Desperate measures

When a region has very little water of its own, one solution is to import (bring in) fresh water from somewhere else. Water exports by tankers may become more common in the future. The downside is that transport causes air pollution, uses energy, and allows people to keep on using water that they should perhaps be saving.

In many parts of the world, essential water supplies have to be transported from place to place in water tanks, rather than by pipes.

In the future, water-poor countries may use ships to drag icebergs south from the Arctic to warmer seas. However, apart from damaging Arctic habitats, scientists are uncertain how this could affect ocean currents and global climate.

CASE STUDY

Buried treasure

On a national scale, the United States has plenty of water, but in certain regions, groundwater reserves are being depleted. Wisconsin is a state rich in water. It has 15,000 lakes and enough groundwater to submerge the whole state 30 m (100 ft) in water! However, this buried treasure is under threat.

In growing settlements such as Madison and Milwaukee, aquifers are being pumped out faster than they can refill. Groundwater levels have dropped by as much as 150 m (450 ft). Wells have run dry and some freshwater springs and streams have stopped flowing. State-wide there is concern for the Great Lakes because less groundwater is topping them up. Wisconsin residents are becoming much more aware of the connection between water resources and human activities.

A boatman checks the water levels at a boat launching ramp on Lake Michigan, USA, in 2007. With falling lake levels, the ramp needs to be extended by an additional 12 metres.

Using technology

Factories can remove salt from seawater in a process called **desalination**. Desalination meets around 60 percent of freshwater needs in Saudi Arabia and other Gulf states, and is important in dry countries such as Australia and Spain. Desalination works in one of two ways. The "reverse osmosis" method uses pumps powered by electricity or by diesel motors. The pumps force seawater through a membrane (a very fine filter) that traps the salt. Another method is **distillation**, where seawater is boiled so that it evaporates into steam. It then condenses and forms drops of fresh water.

Disadvantages to desalination

Reverse osmosis equipment involves expensive technology and all desalination uses lots of energy. Nonetheless, many places where desalination is required are sunny, and solar power can be used to meet electricity needs. However, desalination leaves behind a large build-up of salty brine, which is difficult to dispose of.

These piles of salt were removed from seawater during desalination. Some of the salt can be sold to the food industry, but much is wasted.

CUTTING EDGE: MADE-TO-MEASURE CLOUDS

Aerosols in the sky help water vapour to condense. Cloud seeding is a technology that uses this knowledge to create rain. Rockets or planes shoot particles of silver iodide or dry ice into clouds. Inside the clouds, tiny water droplets merge with these aerosols. The droplets are high in the sky and they gradually join with other droplets to form snow. As this snow falls it melts into raindrops. This method of getting extra water is expensive and not without problems. In China, where cloud seeding is used, one region effectively steals rain from a cloud that was due to deposit rain on another region.

CUTTING EDGE: THE SEAWATER GREENHOUSE

To provide a low cost and sustainable means of producing fresh water for crops in hot, dry regions, British scientist Charlie Paton came up with an ingenious solution. He designed a greenhouse that uses the Sun's heat to distill seawater to irrigate crops and enable them to grow in cooler conditions.

Plants need sunlight in order to grow, but too much Sun can kill them. In hot regions, plants die because they lose too much water. In the Seawater Greenhouse, some of the heat is taken out of the sunlight, allowing the plants to grow in cooler, yet high light conditions. This heat (solar energy) is then used to distill fresh water from seawater. Paton says of his cost efficient and environmentally friendly invention, "In theory our simple technology could turn the Sahara back into a forest, as it was 6,000 years ago."

Using less

There are many ways in which individuals and governments can conserve water. The agricultural industry conserves water by improving irrigation methods and growing drought-resistant or less thirsty crops. Manufacturing industries are developing machines and technology that use less water. For example, the steel industry uses water to cool hot metal. Installing technology to re-use coolant water is one reason the industry now uses 25 percent less water. In cities, governments and householders can save significant amounts of water by mending leaky taps. One drip per second wastes around 1,200 litres (2,540 pts) of clean water each year. In rural areas in LEDCs, lining soil wells with impermeable concrete can help prevent water loss.

Harvesting and greywater

Collecting (harvesting) rainwater and recycling washing water, known as greywater, are important principles of water conservation. People collect rainwater by diverting rain falling on roofs and gutters through drainpipes and into water butts or tanks. Harvesting saves water and money. Rainwater is free and easily cleaned by sieving to remove bugs and leaves. Architects are designing new houses and other buildings that divert greywater from sinks, baths, showers, and washing machines to fill up cisterns.

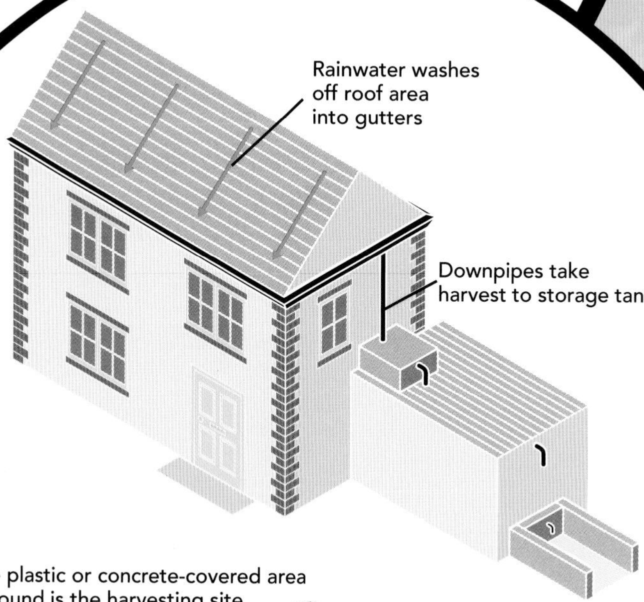

Rainwater washes off roof area into gutters

Downpipes take harvest to storage tank

These diagrams show two simple rainwater harvesting systems that can be used in homes.

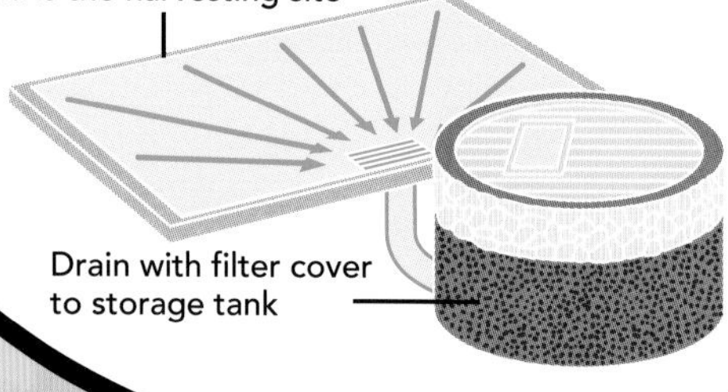

Large plastic or concrete-covered area on ground is the harvesting site

Drain with filter cover to storage tank

Mexican solutions

Mexico City was built on the site of an old lake. The city does not have access to a surface water source, so its 20 million inhabitants rely on an underground aquifer. Over the years, as more and more water has been pumped out from beneath the city, the land has been subsiding (sinking). It is estimated that in the last 100 years, the city has sunk more than 9 m (30 ft), leaving cracked pavements and roads, buckled walls, and buildings leaning at strange angles.

In 2005, Ilan Adler, an environmental scientist at Mexico City University, set up large black plastic tanks to harvest rainwater. They collect enough water to provide around 80 percent of the water used by the university's teachers and students. Today, the city's government is looking at ways of expanding the wastewater network, and funding pipe repairs that will reduce wastage. Their focus is on managing water demand through water pricing and metering, educating people in the city about the need for water conservation, and wastewater collection and re-use programmes.

Sustainable water use

According to the United Nations (UN), more than 60 percent of the world's population will be suffering from water shortages by 2025 if population and water consumption rates continue to rise at current levels. Even in continents with managed water supplies and relatively damp climates, such as Europe and North America, there are increasing periods of drought.

In order to provide clean water and sanitation, and to conserve fresh water for future populations, water supplies must be managed carefully. Stricter rules and regulations can reduce pollution. Educating people in MEDCs about water issues can encourage them to use less and recycle more water. Providing poorer people with the means to improve their own water and sanitation supplies can make a real difference.

Facts and figures

Timeline

800 BC First sewers built in Rome.
300 BC First Roman aqueduct is built.
260 BC First reservoir is built in Rome.
200 BC Irrigation by a device called a shaduf is first carried out.
1340s The Black Death (plague) kills millions across Europe, spread by lack of sanitation and dirty water.
1400s Most sewers in European cities are streams or open gullies running along the streets.
1510 First underground sewer is built in London.
1665 Plague kills 60,000 people in London.
1775 Alexander Cummings, a British watchmaker, takes out the first patent for a water closet (flushing toilet).
1800s To improve sanitation, many fountains and urinals are built in Paris, France.
1800 Chlorine is first used as a water purifier.
1829 Scientist John Simpson designs a slow sand filter for drinking water.
1832 Cholera kills millions across Europe and the United States.
1854 British surgeon, John Snow, proves cholera is spread by water contaminated with faeces.
1860s London and other major European cities are still full of open sewers.
1870s Across the world, human excrement is used as a fertilizer and spread on fields.
1880 Melbourne, Australia, is overcrowded and a nearby river is an open sewer.
1902 First permanent water chlorination plant is built in Belgium.
1914 First drinking water standards are set out by the United States.
1925 First hydroelectric dams in the United States pump irrigation water.
1930s Wide-scale use of chlorination of water sees cholera death toll drop dramatically.
1936 Nearly 60 percent of all farm homes in the United States have flowing indoor water.
1945 Fluoride is first added to drinking water supplies in the United States to reduce tooth decay.

1950s Armed conflict between Israel, Jordan, and Syria arises over use of River Jordan water. Kuwait becomes the first state in the Middle East to use seawater desalination technology.

1997 China convicts someone for water pollution for the first time.

1998 Water contaminated by arsenic affects 25 million people in India.

1999 Ultraviolet light is found to be an efficient method of disinfecting wastewater before it is released into rivers and lakes.

2000 Drought in Asia affects 60 million people.

2000 In more than 50 percent of LEDCs, less than half the population has sanitation or a source of clean water.

2001 Due to global warming, the ice cap at the North Pole is 40 percent thinner than it was in 1950.

2006 Countries in the European Union all have urban sewage collection and treatment facilities.

2008 Technology is used to recycle wastewater from sewage into drinking water in many areas of the world.

2008 In parts of the world, people still walk many kilometres every day to collect water.

2025 Nations are expected to face critical water shortages.

Global freshwater distribution

The table below shows how much fresh water was present in the different regions of Earth in 2002 (measured in cubic kilometres).

Region	Glaciers and icecaps	Groundwater	Wetlands, lakes, rivers, and reservoirs
North America	90,000	4,300,000	27,000
South America	900	3,000,000	3,400
Europe	18,200	1,600,000	2,500
Africa	0.2	5,500,000	31,800
Asia	61,000	7,800,000	30,600
Australasia	180	1,200,000	221
Antarctica	30,109,800	—	—
Greenland	2,600,000	—	—

The column on the left shows the percentages of fresh water and salt water on our planet. The column in the centre reveals the small percentage of fresh water that is found on the surface. The column on the right shows that most of our surface water is found in lakes and reservoirs, with very little in rivers.

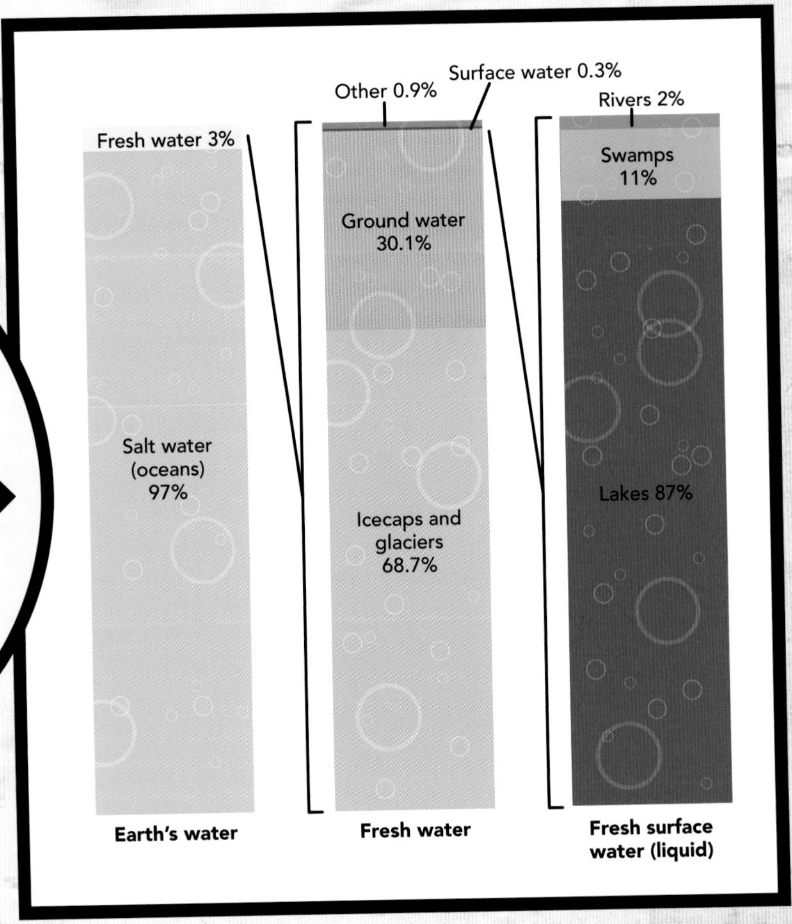

Water usage around the world

- People who do not have running water in their house or garden, but who live within one km (0.6 miles) of a water source, use approximately 20 litres (42 pts) of water each day. This includes 1.8 billion people worldwide.
- In some MEDCs, the average person uses more than 50 litres (106 pts) of water each day just to flush toilets.
- The highest average water use in the world is in the United States where each person uses 600 litres (1,270 pts) of water each day.

Amounts of water needed to make different products

- It takes 2.5 litres (5.3 pts) of water to produce one litre (2.1 pts) of petrol.
- It takes 7 litres (14.8 pts) of water to make a 100-g (3.5-oz) bar of chocolate.
- It takes 9 litres (19 pts) of water to make one magazine.
- It takes about 2,700 litres (5,706 pts) of water to make one cotton T-shirt.
- It takes up to 4,000 litres (8,454 pts) of water to produce one kg (2.2 lbs) of wheat, and up to 16,000 litres (33,810 pts) of water to produce one kg of beef.
- It takes one million litres (2.1 million pts) of water to make one tonne (1.1 tons) of steel.

Find out more

Further reading

At Issue: Will the World Run Out of Fresh Water?, Debra A Miller (Greenhaven Press, 2007)

Energy Debates: The Pros and Cons of Water Power, Louise and Richard Spilsbury (Wayland, 2007)

Experimenting with Water, Robert Gardner (Dover Publications Inc., 2004)

Natural Resources: Water and Atmosphere, Julie Kerr Casper (Chelsea House Publishers, 2007)

Our Fragile Planet: Hydrosphere: Fresh Water Systems and Pollution, Dana Desonie (Chelsea House Publishers, 2008)

The Atlas of Water: Mapping the World's Most Critical Resource, Robin Clarke and Janet King (Earthscan, 2004)

Websites

www.bbc.co.uk/weather/world/city_guides/
Take a look at the average monthly rainfall for cities around the world.

www.guardian.co.uk/environment/water
Keep up to date with water issues and news.

www.oxfam.org.uk/coolplanet/kidsweb/oxfam/unwrapped/safe_water.htm
Find out how Oxfam helps after water disasters (such as the Indian Ocean tsunami in 2004).

www.wateraid.org.uk
Find out about the charity WaterAid and how they try to give the world's poorest people access to safe water.

Research topics

- Find out more about the bioaccumulation of DDT in birds of prey. What happened to the birds' eggs, and what has been done about the problem? You could start your research at www.chem.duke.edu/~jds/cruise_chem/pest/pest1.html
- Learn all about new desalinating plants. Are there plans for a plant in your country? Why is the issue of energy use causing some countries to block new desalinating plants?
- Research how women and children are affected by limited water supplies. How much schooling do children miss and how does this affect their future work prospects? You can find information about the links between education and sanitation in India at http://www.lboro.ac.uk/well/resources/Publications/Country%20Notes/CN2.2%20India.htm

Glossary

acid rain rainwater that has been polluted by chemicals in the air, making it acidic and damaging to wildlife

aerosol tiny solid or liquid particle (other than water or ice) suspended in the atmosphere

aquifer underground layer of sand and rock that is saturated with water so that it acts as a water source for a well or spring

atom one of the tiny particles that make up matter

bioaccumulation process by which the concentration of toxic chemicals increases in living tissue, such as in plants or fish, as they breathe contaminated air, drink contaminated water, or eat contaminated food

condense change from a gas to a liquid

conduction movement of heat or electrical energy from one material to another by direct contact of one molecule to the next

desalination process in which dissolved salts are removed from seawater to make it drinkable

distillation process in which one substance is boiled away from another as the heat turns it into a gas or vapour. It is then collected and cooled back into a separate liquid.

drainage basin total area of land from which water drains into a single body of water such as a river

erosion when rock or soil is worn away by wind or water

eutrophication process by which large additions of nutrients to a body of water causes a rapid increase in algae and a depletion of oxygen

evaporate change from a liquid to a gas

evapotranspiration combination of evaporation of water from the soil and the transpiration of water by the plants that live in that soil

food web diagram that shows what eats what in a habitat

fossil fuel fuel formed from the remains of plants and animals that died millions of years ago, for example, coal, oil, and gas

geoscientist person who gathers and interprets data about the Earth and other planets

hard water water that contains a high concentration of dissolved minerals such as calcium and magnesium

humidity moisture in the atmosphere

hydrological cycle movement of water between the ocean, the atmosphere, the land, back to the ocean, and so on

impermeable type of material through which fluids cannot pass

infiltration movement of water from the land surface into the soil

irrigate supply water from artificial channels or ditches, pipes, or streams to grow crops and other plants

LEDC Less Economically Developed Country, a country where many people are poor and do not have basic services like clean water or schools

longshore drift when waves drag sand and pebbles in a zigzag movement along a beach

MEDC More Economically Developed Country, a country where the average income is relatively high and most people have access to services such as schools and hospitals

molecule two or more atoms joined together

parasite animal or plant that lives in or on another animal or plant, and takes nutrients from them

permeable type of material through which fluids can pass

photosynthesis process by which plants make food in their leaves using water, carbon dioxide from the air, and energy from sunlight

precipitation water that falls to Earth in the form of rain, snow, hail, or sleet

renewable something that can be replaced

resource material that people use such as water, soil, air, trees, and fossil fuels

saturated completely full

sediment tiny pieces of rock and mud that settle at the bottom of a river

silt sand or mud deposited by flowing water

soft water water that contains very few, or no, dissolved minerals

solute substance that dissolves

solution mixture of dissolved substances

solvent substance in which a solute dissolves

spit strip of land across a bay or river mouth

water table top level of water stored underground

weathering process in which rocks that are exposed to the weather are worn down by water, wind, or ice

Index